Tupac, Biggie & Dinosaur Killing Asteroids

By Gary C. Booker

ISBN: 978-1-008-94401-5

Table of Contents

Chapter One: Primordial Soups, the 1970's and New York......................4
Chapter Two: Stromatolites and Superstars..12
Chapter Three: Tupac, Biggie, and Dinosaur Killing Asteroids..............24
Chapter Four: Iced Out Continents, Iced Out Chains.............................38
Recommended Reading...51

This book doesn't have a preface. Just follow along. Welcome to the inside of my head. Wipe your feet before you come in.

Chapter One: Primordial Soups, the 1970's and New York

Scientists use the term "primordial soup" to describe the part of the Earth's early history where the presence of complex organic molecules, water and random geological processes were just right for the earliest and simplest forms of life to emerge. Archaea, the domain that composes the earliest living organisms, appeared about 3.7 billion years ago, which is relatively early in the Earth's history. They appeared near hydrothermal vents in the Earth's oceans. To this day, archaea can be found in the most extreme environments that would obliterate any other living organism.

Early Earth was a hostile place. It was nothing like the pale blue dot that we call "Mother Earth" today. The amenities of an ozone layer, oxygen rich atmosphere, and relatively warm climates are things that came much later, all making it possible for life on Earth as we know it to exist. Earth began as a hell of molten rock newly formed by the accretion of particles around its newly formed star. A collision with another protoplanet resulted in the formation of the moon. Years of cooling and the release of water vapor from rocks would result in extended rainfall that would form our oceans. The crude atmosphere of early Earth, which contained mostly carbon dioxide along with methane, ammonia, and traces of water vapor and very small traces of oxygen, would kill most of today's living things within a matter of seconds. The surface temperatures were extremely hot, beginning at temperatures well over a thousand kelvin and cooling over time very slowly to temperatures much cooler but still far too hot for most life today.

But these rough, inhospitable conditions are where the first microorganisms got their start. They weren't as glorious as the trilobites of the Paleozoic, the dinosaurs of the Mesozoic, or the wooly mammoths, Sabers toothed tigers, and humans of the Cenozoic. But they are how this whole life thing got started, and we have to give them the same props that we give to Clive Campbell (DJ Kool Herc), Joseph Saddler (Grandmaster Flash), and the other talented musicians that created hip hop under the challenging realities of living in 1970's New York City, which was its own primordial soup of sorts.

New York City is divided into 5 Boroughs. The Bronx was a place that saw a mixture of Black American, Caribbean, and Latino/Hispanic cultures that all brought their own musical traditions to a vibrant party scene. New York City was plagued by drugs, gangs, urban blight, pollution during the 1970's. The gangs included the Black Spades, the Spanish Cobras, Ghetto Brothers, Roman Kings, Screaming Phantoms, and many other highly organized street organizations with heads, mid-level bosses and street enforcers. There was a considerable amount of violence during these times that overlapped with the party and music scene.

Furthermore, the economic woes of the time were exacerbated by an Organization of Petroleum Export Countries (OPEC) embargo of the United States and several other countries in 1973. Secretary of State Henry Kissinger was able to negotiate the end of the embargo in 1974. But the effects of the embargo on the prices of oil would persist throughout the decade, affecting the entire economy.

Lance Taylor was a young Black Spade in the 1970's. But Taylor was no ordinary Spade. He had aunts and uncles that were participants in Black Liberation Movements that started to catch Taylor's ear. Taylor was also a skilled writer and won a trip to Africa in an essay contest. His trip to Africa, along with inspiration from the movie Zulu, inspired him into changing his name to Afrika Bambaata. Bambaata then set out to use music parties to unite gangs into what became known as the Universal Zulu Nation.

These parties would invite some of the best known DJ's at the time. Masters of ceremonies, known as MC's, would lead the parties. Their skills of facilitation would draw upon musical traditions of various soul singers, poets, and other artistic traditions of the time. Then, there were dancers. Break dancing, or breakin', was very popular and there were crews of breakers with various different gangs. These parties gave the local gangs a new way to settle their differences. Rap battles, dance battles, and DJing contests became alternatives to knives, guns and baseball bats.

We began to see what Temple of Hip Hop founder and Golden Age emcee (MC) Lawrence "Kris" Parker (KRS ONE) began to refer to as hip hop's 9 elements:

1 Breaking
2 Emceeing
3 DJing
4 Grafitti
5 Beat boxing
6 Street Fashion

7 Street Knowledge

8 Street Language

9 Street Entrepreneurship

The Hip Hop DJ transforms the record player from an ordinary sound playing device into a bona fide musical instrument. Beginning with Grandmaster Flash, the Hip Hop DJ takes samples from multiple records, using precisely timed transitions and sound mixers to produce new music. The beat boxer becomes the musical instrument, making a variety of different musical sounds with his/her mouth/body. The economy, coupled with the massive teacher layoffs, and other issues that played New York City's Department of Education beginning in 1968 made it difficult for many Bronx youth to have access to music teachers. Hip hop became a result of Bronx youth making do with what was available.

Archaea are single celled organisms. They are prokaryotes: organisms with cells that do not have a nucleus. The genetic material of these organisms is located in the center, which is known as the nucleoid region. We see cell walls, cell membranes, ribosomes, cytoplasms, and other things that make up a living cell. Flagella and pili allowed for locomotion. These crude organisms met the conditions of all living organisms:

1 Metabolism
2 Reproduction
3 Response to Stimuli
4 Grow and Change
5 Maintain homeostasis
6 Composed of cells

There are multiple things that likely contributed to the formation of life on Earth. It's location in our sun's inhabitable zone, it's composure of carbon based compounds, the magnetosphere to limit the levels of radiation on the Earth's surface, and the deposits of water and other essential chemical compounds from meteor, asteroid and comet impacts during the Late Heavy Bombardment (4.5 to 3.8 billion years ago). It is possible that the first microorganisms on Earth originated from elsewhere and traveled here via one of the aforementioned impact objects.

Endosymbiosis is the phenomenon where one organism begins to live inside of the other. This played an important role in the evolution of the eukaryotic cell, or cells that have a nucleus. The mitochondria in our cells is a result of this. The nucleus would evolve as the result of a need for better transcription of genetic material. The jury is still out on how this occurred. Protists were the earliest eukaryotes and first appeared around 2 billion years ago.

Hip hop's own endosymbiosis was the joining different, independently existing lifestyles into one. Breaking, beatboxing, djing, graffiti art, and MCing were all separate things that existed independently before hip hop brought them together into one culture. Being inside of one culture caused each activity to co-evolve with the others. The hip hop DJ became drastically different from the DJ's at other parties with an incredibly different critical skill set to learn. The MC became more than just a master of ceremonies and developed a skill set that was a cross between a singer and a poet, which gave the phrase "rapping" (originally black baby boomer slang for going on a tirade) a new meaning associated with hip hop such that the terms "rapper" and MC became interchangeable.

Speciation occurs when organisms evolve so much over time to the point where they are no longer the same species as before. These changes are gradual and occur over time. The gradual changes in prokaryotic organisms would lead to the emergence of Eukaryotes. Similarly, "hip hop" became a thing when the observer was no longer looking at something that was distinctly R&B, Rock, Reggae, Funk, Disco, or anything else.

The earth was too hostile to evolve complex multicellular life. Despite simple unicellular life forms appearing early in the Earth's history, a variety of geological and atmospheric changes would need to happen before complex lifeforms would begin to appear. Similarly, a variety of different things would need to occur in order for hip hop to become the multi-billion dollar juggernaut filled with world famous superstars and pop culture icons that it is today.

Chapter Two: Stromatolites and Superstars

Stromatolites are layers of rock that date back to the time period known as the Precambrian. They were a very important step forward in the Earth's history. They appeared when the Earth was still inhospitable to multicellular life. There were many unicellular life forms around during this time. But stromatolites contained one such unicellular organism that would stand out from the rest in a way that would completely change the trajectory of life on Earth: Cyanobacteria, or "Blue-Green Algae."

Cyanobacteria had a special talent: photosynthesis. The combination of water and carbon dioxide, which was abundant in its surroundings at the time, and transforming it into oxygen and glucose. These little creatures would have an enormous impact on the planet by filling the atmosphere with oxygen, causing a reduction in global temperatures. Many of the early bacteria species would be wiped out by the changes in oxygen levels, resulting in the Earth's first mass extinction. Oxygen levels would tremendously rise over a period of billions of years, and the atmosphere would cool so much that the oceans would begin to glaciate and cover the entire Earth, beginning the time period known as "Snowball Earth."

The late 70's arrived, the years of the Jimmy Carter Administration were coming to an end, and the 1980's were approaching. Hip hop had grown into a regional cultural phenomenon similar to that of go-go music in Washington DC (which would later influence hip hop), but it had yet to break into the mainstream. Enter Robert Ford, Jr., a music journalist and record producer. Ford was a journalist for the newly formed Billboard magazine and was instrumental in the documentation of hip hop culture. He would later meet Russel Simmons, who introduced him to Kurtis Walker. Walker was an MC who went by the name of Kurtis Blow. Ford was interested in songwriting and wrote "Christmas Rappin" for Kurtis Blow, which was released in 1979. The single was a surprise hit. It led to Mercury Records signing Kurtis Blow to a record deal, where Kurtis Blow would release a self-titled album.

Kurtis Blow would become the first rapper to become an artist on a major record label and subsequently the first commercially successful rapper. It was a major precedent, as Kurtis Blow proved to music insiders that were initially dismissive of the commercial viability of hip hop that the artform could sell records. Other acts would follow. Grandmaster Flash and the Furious Five, Funky Four Plus One More, the Sugar Hill Gang, Fab Five Freddy, the Treacherous Three, and other artists would release albums and see commercial success. Many of these early Albums were influenced by disco, which was enormously popular back in the 1970's. The disco audience also took to hip hop. This was both a blessing and a curse.

American politics have always been ripe with culture wars. And disco became the new punching bag for the American right to blame for just about everything it deemed to be a social problem. Major League Baseball had hosted "Disco Demolition Night" in 1979, where several records were destroyed in front of a crowd. The night ended in a riot. It was a major catalyst in anti-disco sentiment that would kill the music genre. It sucked the life out of early commercially produced hip-hop. But things were far from over. The hip hop scene was only continuing to evolve. Russell Simmons, influenced by his involvement with the early success of Kurtis Blow, would begin making preparations to start hip hop's own record label. Russell was one of many highly resourceful visionaries that would navigate the various obstacles of the 1980's to give rise to hip hop's golden age.

Resourcefulness could also apply to life in the oceans during Snowball Earth, when glaciers reflected the majority of the sunlight back into space. Aerobic life forms needed oxygen to survive, and the oxygen cycle relies on photosynthetic organisms. Volcanoes would save the day. Glacial meltwater, or water produced by the geothermal melting of glaciers, would continue to supply aerobic life with its much needed oxygen supply during the frozen periods of Snowball Earth. Those organisms continued to evolve into more complex organisms over time. Eukaryotic organisms grow more sophisticated in their environments.

Extended periods of major volcanic eruptions, releasing large amounts of carbon dioxide in addition to the heat of its eruptions, would result in the breakup of glaciers and the reemergence of the Earth's crust. The temperate climate that we now know eventually emerges, and the Earth enters a time period known as the Cambrian Explosion. The Cambrian Period is the first period of the Paleozoic Era. It is during this time period that we see an explosion of lifeforms in every kingdom that we know today: Archaea, Bacteria, Protists,

Fungi, Plants and Animals. Obabinias, trilobites, anomalocarises, are examples of prehistoric arthropods seen in the oceans. Conodonts and metasprigginas are some of the first fish. The Cambrian explosion happened in the Earth's oceans, but the time period also marks the beginning of the formation of the Earth's Ozone layer, which would drastically reduce the amount of ultraviolet radiation on the Earth's surface and make terrestrial life possible.

The "New School" of the mid 1980's brought their own sound. It wasn't the disco-esque sound of Kurtis Blow and the Sugar Hill Gang. It made use of the newly invented beat machine. The lyrics became more sophisticated, aggressive, and commentary based, as seen in "The Bridge" by Shawn Moltke (MC Shan) of the Juice Crew. Artists like Lawrence Parker (KRS-ONE) and Scott Sterling (DJ Scott La Rock) of Boogie Down Productions sampled songs from James Brown, who embraced hip hop. Artists like Run DMC embraced rock and roll, and many of their songs would blur the line between hip hop and heavy metal. Their collaboration with Aerosmith on "Walk This Way" would influence both hip hop and rock and roll in the years to come.

Hip Hop remained competitive. There were various different artists feuds that resulted in "battles" on wax. There were the “Bridge Wars” between KRS One & Boogie Down Productions and MC Shan & the Juice Crew. There were the “Roxanne Wars,” between several artists, including Roxanne Shante of the Juice Crew. Mohandas Dewese (Kool Moe Dee), formerly of the Treacherous Three, feuded with Todd James Smith (LL Cool J). Kool Moe Dee is possibly the inventor of the modern diss track. Prior rap battles were contests between MC’s to see who could engage the crowd the most. Kool Moe Dee was the first rapper to actually freestyle lyrics that verbally attacked the other contestant in a battle with David James Parker (Beatrocker Busy Bee Starski).

Hip Hop’s New School era was a lot like the Cambrian explosion, where we would see the emergence of several rappers that would go on to make classic hits. In the mid-80’s we began to see the emergence of sub-genres. Gangsta Rap started with Jesse Bonds Weaver (Schooly D) in Philadelphia and was also influenced by Boogie Down Productions’ “Criminal Minded” album. Boogie Down Productions would also influence conscious rap with it’s more politically charged lyrics on it’s next album “By All Means Necessary,” and it’s collaboration with other East Coast rappers under the name Stop the Violence Movement to record the song “Self

Destruction" (a response to the tragic killing of Scott La Rock in 1987).

The year 1987 saw the entry of very talented rap superstars that would be so influential that they would influence every artist to come after them, beginning with William M. Griffin, Jr. (Rakim). Rakim was a student athlete in high school with his eyes on playing in college. But he was also a trained saxophonist with parents that were also trained musicians. He grew up around hip hop and was influenced by the rising stars of the 1970's and early 1980's. Rakim developed a style of rapping that was influenced by the improvisation of John Coltrane. After meeting Louis Eric Barrier (Eric B), he went into the recording studio and made "Paid in Full."

The sound was so different that Marlon Williams (Marley Marl), who was producing the song, initially clashed with Rakim. Marley Marl thought that Rakim's flow lacked energy and suspected that Rakim lacked motivation. After Eric B insisted on letting Rakim record the record as he pleased, Rakim finished the song and it was released. The song was an instant hit. But it was more than a hit. It raised the fanbase's expectations of lyrical sophistication so much that it ended the careers of artists that couldn't keep up, and paved the way for similar artists. Along with Rakim and KRS ONE, we see Richard Martin Lloyd Walters (Slick Rick), Antonio Hardy (Big Daddy Kane), Lana Moorer (MC Lyte), Dana Owens (Queen Latifah), and other artists that have lyrical skills that are far ahead of the first generation of new school rappers.

The Ordovician-Silurian (O-S) Extinction was the first mass extinction of the fossil record. Smaller extinctions appear sporadically throughout history. But a mass extinction occurs when several species die at once due to a common trigger or set of triggers. The triggers of the O-S are a disruption of the global carbon cycle due to volcanic activity, and the drastic climate changes that became as a result. About 85 percent of life would die out. But the Silurian period would see lifeforms conquer the land. Plants had made landfall during the Ordovician period after the newly formed ozone layer eliminated deadly levels of radiation. The first land animals would begin to make landfall during the Silurian period.

The Silurian, Devonian and Permian Periods would see an evolving Earth that would become more and more like what we would expect today. Fish, amphibians, reptiles, arthropods, echinoderms, and a variety of different organisms would populate the Earth. The continents were forming the supercontinent that paleontologists refer to as Pangea, which would be completely formed from previous continents by the Permian Period. Today's organisms are descendants of organisms that lived during this time.

Hip Hop's golden age is considered to have been from about 1986 to 1990 by many music historians. And like with the paleozoic era, we see many rappers that will develop styles that will influence newer generations of rappers for years to come. For example, Rakim's style of rapping would influence rappers like Tupac Shakur (2pac), Nasir Jones (Nas), Marshall Mathers (Eminem), and Kendrick Lamar. Big Daddy Kane's style of rapping would influence rappers like Nathaniel T. Wilson (Kool G Rap), Christopher Wallace (The Notorious B.I.G.), Kimberly Jones (Lil' Kim), Shawn Carter (Jay-Z). Queen Latifah would influence rappers like Lauryn Hill, Melissa Elliot (Missy Elliot), and Anthony Chris (Treach). The aforementioned rappers are not carbon copies of their influencing rappers, but their influence can definitely be heard. As with living organisms, the evolution of music occurs with gradual changes with every artist putting their own twist on the genre in which they contribute.

Rap Groups like Public Enemy and A Tribe Called Quest would set the tone for later rap groups to follow. William

Drayton (Flava Flav) of Public Enemy would set the tone for hype men in rap groups. We would see his influence in Wu Tang Clan, Lost Boyz, and other groups.

By the late 1980's, there were radio stations that specialize in hip hop springing up in every major city. Rap music was appearing in cartoons. The Cable Network Music Television (MTV) had begun playing rap videos and started the segment "Yo! MTV Raps" in 1987. The West Coast had emerged with its own hip hop scene, where Tracy Marrow (Ice-T), Todd Shaw (Too $hort), and rap groups like JJ Fad and NWA were breaking out and getting national airplay.

Lots of people wanted in, one way or another. And this included lawyers for companies that took up issues with hip hop's use of sampling. A major court case became inevitable. And Grand Upright Music vs. Warner Brothers Records was a dispute between Gilbert O. Sullivan and Marcell Hall (Biz Markie) over Sullivan's song "Alone Again," which Biz Markie sampled. The court ruled in favor of Sullivan. The ruling required artists to clear every sample before releasing it. This killed hip hop's first golden age, which heavily relied upon song sampling.

The late Permian Period saw the development of the supercontinent Pangea. The Earth has always been volcanically active, but the presence of a supercontinent has created a unique set of dangers that could spell disaster for living organisms in the event of a volcanic eruption. Pangea was so large that it is impossible for rain to reach it's interior, creating an enormous desert in the continent's interior. A supervolcano, located at the present day Siberian Traps, erupted for two million years, killing the majority of all life on Earth, ending the Permian period and the Paleozoic Era and beginning the Triassic Period & Mesozoic Era. The Permian-Triassic (P-T) Extinction, also known as "The Great Dying," is the bloodiest mass extinction on the fossil record.

Just as the P-T Extinction nearly eradicated everything in the Paleozoic, the aftermath of Grand Upright Music vs. Warner Brothers Records would reverberate throughout the music industry, as the sampling traditions of hip hop became too cost prohibitive to continue. Entire record albums had to be put on the shelf because samples could not be cleared, forcing record labels and artists to eat the enormous costs of studio time with zero profit. The careers of many hip hop artists as well as producers would not be able to adapt.

But life on Earth went on, just as hip hop continued. The Mesozoic Era had begun, which would compose of the Triassic, Jurassic and Cretaceous Periods. Pangea began to break up into the supercontinents of Laurasia and Gondwanna during the Triassic Period, changing the climate patterns of the land masses. The first dinosaurs evolved from surviving reptiles. These reptiles would continue to evolve in ways that would eventually produce mammals and birds.

Ice-T would continue to do well as a rap artist until the mid 1990's, but also added Rap Metal to his repertoire and formed the rock band Body Count. Ice-T would also shift to acting, starring in the movie New Jack City with Judd Nelson, Wesley Snipes and Mario Van Peebles. LL Cool J continued to be successful as a rap artist and would also venture into acting. But most of the rappers from their eras would fade to black, become DJ's (Biz Markie was one of many rappers that would transition to DJing) or start new careers as actors (Will Smith, Queen Latifah, and others would follow suit). Meanwhile, a new generation of artists were in the works from different parts of the country that were now developing their own hip hop scenes that were starting to get national attention from record executives looking for the next big thing. A second hip hop golden age was coming.

Chapter Three: Tupac, Biggie, and Dinosaur Killing Asteroids

I wasn't around for the extinction of the dinosaurs. That was about 65 million years before my time. But I know from my studies of paleontology that a large asteroid struck the planet, landing in what is now the Chicxulub crater in the Yucatan Peninsula. The energy of the impact was equivalent to 10 billion of the atomic bombs that were dropped on the cities of Hiroshima and Nagasaki, Japan. The sheer force of the impact, in addition to the enormous dust and debris that were launched into the atmosphere to block the sun, and the hypercanes that were generated by the extreme energy levels in the ocean were enough to kill everything that weighed more than 6 lbs. This completely changed the fossil record.

The tiny mammals that were able to thrive under these circumstances would evolve over millions of years to dominate the earth as rhinos, elephants, whales, and primates... which would eventually include 7 billion ozone depleting, deforesting, asbestos inhaling, acid rain making, greenhouse gassy ass Homo Sapien Sapiens, who would do things that would require impressive amounts of combined intelligence and stupidity like collectively destroy animal habitats to increase the chances of zoonotic viruses like SARS-COV-2 to enter the human ecosystem and then engage in xenophobic pissing contests over who's to blame for spreading it. Then, there are times where these humans would also make music and then fight each other (and occasionally kill each other) over the music.

It seems like a weird comparison, but it reminds me of the tragic murders of Tupac Shakur (Who used "2Pac and "Makaveli" as his stage/album names) and Christopher Wallace (Known by his stage name "The Notorious B.I.G.," and sometimes "Biggie Smalls" and "Big Poppa"). These two titans dominated the worlds of hip hop just as Tyrannosaurus Rex and Triceratops dominated the Cretaceous period. I was a 9th grader at Daniel McLaughlin Therrell High School in Atlanta, Georgia at this time and remember it well. And just as the fossil record was instantaneously transformed by the Cretaceous-Palogene (K-Pg) Asteroid Impact, the world of hip hop would never be the same after the worlds of the deaths of Pac and Big.

The year 1996 was marked by a bicoastal rap feud that was growing increasingly intense. Hip hop as we know it has its origins in New York City. South Bronx, to be exact. The musical elements that are responsible for the creation of hip hop come from just about everywhere that Black people roamed the earth. The American south, midwest, west coast, the Caribbean, Latin America, West Africa, you name it. The Bronx, with all of it's diversity, was the perfect place for this to start. It would eventually spread throughout the 5 Boroughs of New York City and then all over the country. Musicians in other parts of the country would begin making their own versions of it, but the majority of the nascent hip hop scenes would remain local. But one scene managed to break through: California. Specifically, Los Angeles and the Bay Area (Oakland, Palo Alto, etc).

Tracy Marrow (Ice-T) was the West Coast's first major act, followed by Todd Anthony Shaw (Too $hort). Long before Ice-T was a mainstay on Law and Order SVU, he was influenced by Philidelphia Rapper Jessy B. Weaver Jr.'s (Schooly-D) pioneering hit "PSK." PSK, released in 1985, was the first gangster rap song. The song heavily influenced Ice-T, who then released his own hit "Six in the Morning." Los Angeles was in the middle of a gang violence epidemic, and home to

America's two most notorious black street gangs: the Crips and the Bloods. This environment provided Los Angeles rappers with plenty of stories to tell. The rap supergroup "Niggaz With Attitude," or NWA for short, composed of members Andre Young (Dr. Dre), Eric Wright (Eazy-E), O'Shea Jackson (Ice Cube), Lorenzo Patterson (MC Ren), Antoine Carraby (DJ Yella), and Kim Renard Nazel (Arabian Prince). NWA, influenced by Ice-T's "Six in the Morning," released their own hit "Boyz in the Hood." The West Coast soon produced JJ Fad, Tone Loc, King Tee, Above The Law, Young MC, DJ Quik, MC Eiht, The Digital Underground (Where Tupac got his start as a roadie), and eventually MC Hammer (who would become hip hop's first megastar).

By 1990, the West Coast was a powerhouse and acquired a large share of hip hop's larger audience across the country and was just getting started. This did not go unnoticed by New York, and several producers and DJ's there began to show some hostility towards the rising west coast acts. The hostility was subtle at first, but by 1991 it started to become visible as Bronx rapper Tim Dog released "Fuck Compton." Fuck Compton dissed (disrespected) just about every west coast rapper (and even singer Michel'le) that was making the Billboard charts at the time. The animosity that some New

Yorkers were starting to show the West Coast was becoming visible to the entire country. And several West Coast rappers responded with their own diss tracks. The East-West feud was on, and it was only going to get worse.

There were other hip hop scenes throughout the country that were also emerging, existing while the two coasts were fighting for hip hop dominance. Miami, Atlanta, New Orleans, Houston, Memphis, Detroit, Chicago, and Cleveland had their own hip hop scenes that stood in the shadows of east and west, but slowly gained momentum. By 1994, hip hop had entered a second golden age. It was no longer just a New York sound, and artists were coming from everywhere.

The Mesozoic Era, composed of the Triassic, Jurassic and Cretaceous periods, is commonly called the Age of the Dinosaurs. They emerged as the predominant animals after the bloodiest mass extinction of the fossil record, which was the Perminan-Triassic extinction. But dinosaurs weren't the only organisms to emerge from this era. The Mesozoic era also saw the first mammals and first birds. Both organisms evolved from homeothermic (warm blooded) reptiles.

The first birds emerged from the genus Archaeopteryx in the Jurassic period along with the first mammals. Archaeopteryxes were transitional animals that were somewhere in between non-avian dinosaurs and birds. They had feathers, and other features that we now associate with modern birds. Confuciusornis was about the size of a crow and was the earliest known bird to have a beak. There were plenty of feathered dinosaurs, such as the velociraptor (which is nothing like what was portrayed in the Jurassic Park movies). And while birds inherited their egg-laying from reptiles, there is a difference between reptile eggs and bird eggs that is a result of bird evolution.

The shape that we associate with eggs are bird eggs. But eggs of reptiles usually have a more symmetric shape. They are also more rubbery. The majority of reptiles do not care for their younglings and instead cover them in sand, and leave. The younglings eventually hatch. The eggshells of birds (and crocodilians, who are also living descendants of dinosaurs) are hard. The mothers sit on the eggs in their nests to keep the eggs warm, which eventually hatch. The mothers also guard the nest and the younglings until they are mature enough to leave the nest.

Shells allow eggs to survive away from water. The eggs of fish and amphibians do not have shells and must be laid in water in order to survive until hatching, which allowed animals to move further inland away from water. We first see eggshells in reptiles and their reptile-like amphibian ancestors in the Carboniferous Period. Birds, which are able to cover much larger distances than terrestrial organisms, evolved eggs with harder shells, reserves of water, and yolks with larger amounts of fats. This allowed them to lay eggs in dry, remote areas that were far from water.

Mammals are the descendants of synapsids and are the only surviving group of synapsid organisms. The first mammals began to appear in the Jurassic period. These were burrowing, foraging animals that were mostly nocturnal so that they could avoid being the snacks of larger animals. Such mammals included Juramaia Sinensis, which was a rodent-like animal and is the oldest known "true mammal." Mammals are known for providing milk to their younglings and mostly giving live birth. But not all mammals gave live birth to their young. Other egg-laying mammals such as the pseudotribos also existed at this time. There are even mammals to this day that continue to lay eggs, such as the Platypus and several species of echidnas.

In general, mammals are warm blooded, hairy/furry organisms that give live birth to younglings (viviparity) rather than laying eggs, in addition to feeding them milk. While nobody does viviparity like mammals, we inherited the trait from our mammal-like reptilian ancestors. To this day, there are reptiles such as Jackson's Chameleons and Spitting Cobras that also give live birth. There is even one lizard in the Congo, the Ivan Skink (Trachylepis ivensii), that has a placenta.

Viviparity allows the mother the opportunity to keep the fertilized egg inside of her, eliminating the need of keeping an incubated nest. It also allows the mother to provide food to the developing fetus from her own digestion process. Birds evolved harder egg shells, but mammalian evolution did the reverse and did away with eggshells and replacing them with a placenta. But the purpose was the same, which was to allow versatility in locations for youngling birth.

The Story of hip hop in the "Dirty South" and the Midwest is similar to the evolution of mammals and birds, and the eventual rise of mammalian dominance of the Cenozoic Era. My hometown of Atlanta produced a sound that evolved from the Miami Bass style of hip hop, which was the first hip hop scene in the South to gain national attention through the likes of Luther Campbell of the 2 Live Crew, along with DJ Magic Mike & MC Madness, Young and Restless, and other Miami acts. Among Atlanta's first regional stars was Andrell Rogers (Kilo Ali). Many music producers and artists began to relocate to Atlanta, a sprawling Black Mecca in the south that offered a booming economy and a lower price of living in comparison to other black metropolitan areas throughout the country, bringing their connections and experience with them. LaFace Records was the first major label to come out of Atlanta, and was the product of Antonio "L.A." Reid, and Kenneth Edmonds (Babyface).

Antwan Patton (Big Boi) and Andre Benjamin (Andre 3000), forming the group OutKast, made their debut album "Southernplaylisticadillacmuzik" in 1994. Miami was still strong, and Houston's Hip Hop Scene was also emerging with Rap-A-Lot Records. Rappers Richard Shaw (Bushwick Bill), Brad Jordan (Scarface), and William Dennis (Willie D) formed the group Geto Boys were enjoying critical success at the time, having already made several hits such as "Mind Playing Tricks on Me" (1991). The south was noticeable, but it still struggled to eclipse the attention of the east and west coasts.

Several hip hop acts outside of New York were making international charts, and OutKast's Southernplaylisticadillacmuzik added Atlanta to the list of locations other than New York that had thriving hip hop scenes. But New York DJ's were not very receptive to the records and were fueling resentment of artists outside of New York. This resentment was apparent at the 1995 The Source

Awards, where attendees jeered every non New York act, including Calvin Boraedus (Snoop Doggy Dogg, later known as Snoop Dogg) and everyone in the Death Row Records Entourage. To make matters worse, Marion "Suge" Knight made an infamous "Come to Death Row" swipe at New York based Bad Boy Records producer Sean Combs (Puffy, later known as Diddy).

Several New Artists were considered for The Source's New Artist of the Year for groups at the 1995 The Source Awards. Among them was Queensbridge's Mobb Deep, composed of Albert Johnson (Prodigy) and Kejuan Waliek Muchita (Havoc). When OutKast won the award, the crowd erupted with booing. As the members of OutKast approached the stage while passing a hostile crowd with artists from the two juggernauts of the industry, Bad Boy Records and Death Row Records, I can imagine the feeling wasn't that different from the mammals of the Cretaceous period navigating in world filled with Tyrannosaurs. Birds and mammals developed warm-bloodedness and their reproductive strategies (bird eggs and mammalian placentas) because they needed to do something different from the dinosaurs in order to survive in a world dominated by dinosaurs.

"Yo, Andre, what you got to say?" asked Big Boi.

"The south had something to say, that's all I got to say," said Andre 3000 to the booing crowd before walking off stage

LaFace Records, No Limit Records, Rap-A-Lot Records, and other labels that were neither east nor west had to develop their own distinct sounds that were neither the mafioso rap flows of The Notorious B.I.G., nor were the distinct West Coast sounds with Dr. Dre beats of 2Pac, Snoop Dogg. They had to be different. They focused on developing their own regional sounds. OutKast's album succeeded in defining the hip hop sound of Atlanta. It was not Boom Bap from New York. It was not G-Funk from Los Angeles. It was not Miami Bass. And eventually, the country wanted more. They began working on their album "ATLiens."

Percy Miller (Master P) was a hard nosed entrepreneur in New Orleans trying to find his break into the music industry. He had grown tired of street life and was honing his hustler instincts into success as a music executive. After years of performing at clubs, selling albums out of his trunk, and other laborious self-promotion efforts, No Limit Records broke into the mainstream and began producing hits like "Bout it." Now, the world began to hear the New Orleans style of hip hop.

There are numerous accounts from journalists, music insiders, rappers, and retired police detectives from the LAPD,

NYPD, and LVPD that have laid out timelines of the events that lead to the tragic murders of 2Pac and Biggie. I'll leave it to readers to look at those accounts to discuss the particulars. The most important thing to note for this comparison is that the majority of the hip hop audience failed to see the deaths of these two men coming. News of 2pac's shooting in Las was initially perceived as just another chaotic event that he would eventually pull out of and write more songs about. There were even initial news reports that reported that he was expected to survive. But Tupac Amaru Shakur was pronounced dead on Friday, September 13th 1996. The shockwaves were instantaneous. Conspiracy theories, rumors, tributes, articles, and statements from various public figures abounded the newspapers and airwaves.

Death Row records would soon crumble apart. Dr. Dre would grow frustrated with Suge Knight's antics and create Aftermath Records. Suge Knight himself would be jailed on a parole violation. Snoop Dogg, the label's next best selling artist after Tupac, would eventually bail for Master P's No Limit Records.

The Notorious BIG would come to perform in Los Angeles in March 1997. It was only six months after the death of 2Pac. And despite several interviews Biggie gave in the area that struck a reconciliatory tone, the visit was seen by many as a disrespectful symbolic dance upon the grave of his former adversary. He would not make it out alive. On March 9th, 1997, Christopher Wallace was gunned down by a driving assailant after he entered a Chevrolet Blazer to leave a party. He was pronounced dead at Cedars-Sinai Medical Center.

Once again, the fallout was swift. Tributes, public statements, and a continued discussion on violence in hip hop were discussed. The void left in hip hop by Shakur's demise was now even larger with Wallace's death. The two largest acts on both coasts,and the gravitational centers of the East-West feud were now dead. Bad Boy records would not decline as quickly as Death Row Records and Sean Combs would continue to grow wealthy as a producer. But the influence of Bad Boy records would never be the same after Biggie's death.

The late 90's and early 2000's saw the rise and eventual dominance of the South, including several best selling albums by OutKast. Master P, Mystical, Mia X, Juvenile, Lil Wayne, Ludacris, Pastor Troy, Goodie Mob, Young Bloodz, Yin Yang Twinz, Trick Daddy, Trina, Gangsta Boo, and several other southern acts would top the charts. Several East Coast artists, such as Jay Z, Busta Rhymes, and many others would record several top selling songs with southern acts. The region would continue to dominate the airwaves until the ubiquity of high speed internet and smartphone technology changed how we accessed music.

Chapter Four: Iced Out Continents, Iced Out Chains

The Year is 1999. Two years since the murder of Christopher Wallace (The Notorious B.I.G.), and three years since the murder of Tupac Shakur. Both the East and the West Coast are still active and producing hit artists. The East Coast had acts like Earl Simmons (DMX), Shawn Carter (Jay-Z), Jeffery Atkins (Ja Rule), Kimberly Jones (Lil' Kim), Inga Marchand (Foxxy Brown), Nasir Jones (Nas), and many others. The West Coast was producing acts such as Ricardo Brown (Kurrupt), Nathaniel Hale (Nate Dogg), and Dedrick Rolison (Mack 10). A few hip hop veterans such as O'Shea Jackson (Ice Cube) and Todd James Smith (LL Cool J) managed to maintain longevity and scored recent hits such as "We Be Clubbin" and "4,3,2,1," respectively. But the South had become the top dog, and a new era was about to begin: the Bling Era.

Atlanta based LaFace Records was in its prime, producing TLC's second album "Fan Mail." *Fan Mail* complete with several hits like "No Scrubs," and "Unpretty." OutKast was putting out it's third consecutive successful album "Aquemini," which featured hits such as the controversial "Rosa Parks," and they were just getting started with what would become a long list of successful albums that would continue well into the 2000's. New Orleans' No Limit Records indeed seemed to have no limit, producing several classic club bangers in the late 90's with Percy Miller's (Master P) "Make em say uhh," and Vyshonn Miller's (Silk the Shocker) "It Ain't My Fault." But another label in New Orleans, Cash Money Records, would release a song that would change the direction of hip hop: Christopher Dorsey's (B.G.) "Bling Bling." "Bling" would be a term for flashy jewelry, which would become the bragging subject of a lot of songs that would be produced during this era. "Iced out" would refer to the jewelry being covered with diamonds.

And speaking of ice... glaciation was a big development in the Cenozoic Era. The Paleogene period was pretty warm, with tropical forests all over the world. Even the poles had such forests. Those rodent-like mammals that were scattering to keep from being dinosaur snacks were now thriving in the Palogene period, and some of them would get larger over time. Reptiles such as prehistoric snakes were also thriving in this period. But carbon dioxide levels began to drop, the world began to cool as a result, and several species died out as a result. The Neogene Period picks up with glaciers forming in the poles, beginning in present day Antarctica and following with the North Pole.

Mammals evolved fur in order to survive in cold places, such as underground burrows. To this day, we humans get goosebumps when it is cold. We inherited this from the ability of our Mesozoic ancestors to stand their hairs up in order to further insulate their bodies. These could come in very useful in a cooling Earth, where the now seven continents would slowly drift towards their present-day positions. A much colder Neogene period gave rise to a generation of large, dominant, furry mammals such as wooly mammoths, mastodons, and saber-toothed cats. The emergence of grasslands would give rise to grazing animals that would eat grass. Ruminant animals (bovines, deer, giraffes, rhinoceroses, etc.) would develop the ability to digest grass with multi-chambered stomachs that have enormous amounts of gut bacteria to break down high-cellulose plant matter. Other animals, such as rabbits and hares, would develop coprophagy: the ability to re-digest fecal matter in order to finish breaking down cellulose.

The first primates appear during the Neogene period and are initially frugivorous tree-dwelling animals. Their initial homes in Eastern Africa were initially dense forest. But climate changes generated by new mountain ranges, drifting continents, and the continued lowering of global temperatures began to transform the forest into grassland. Something had to give, and the primates would have to adapt to their changing surroundings.

Our comparative story this chapter is one of iced out continents and iced out chains. The phrase "bling bling" made a cultural impact that started a jewelry craze. One of Slick Rick's most famous lines is, "Great Scott, are you a thief? It seems that you have a mouth full of gold teeth!" Chains had been around in hip hop since the beginning, and snatching someone's chain was one of the biggest acts of disrespect that you could do to someone in public. One of the pivotal moments that led to the altercation that led to the fatal shooting of Tupac Shakur was the snatching of Travon Lane's Death Row Records chain by Orlando Anderson (who was jumped by Shakur and the Death Row Entourage while in Las Vegas and is widely believed to be Shakur's shooter). But this craze took the obsession with Jewelry as a status symbol to a complete new level. And there was immediate pushback from the section of the hip hop audience that wanted more than flashy jewelry and sex.

"Underground" hip hop saw an increasing audience as a result of the pushback. Ian Bavitz (Aesop Rock), Yasin Bey (Mos Def), Talib Kweli, Lonnie Rashad Lynn (Common), and Michelle Johnson (Meshell Ndegeocello) were among many of the artists that thrived in this audience. Groups like Binary Star and Dead Prez also thrived during this time. Much of the political themes of the music of this time was pushback against the administration of President George W. Bush. By 2007, most of the hip hop audience had grown tired of "bling" oriented music for miscellaneous reasons. One of them was a worsening economy that would eventually turn into one of the worst recessions in recent history.

But the biggest disruptive force in hip hop would become high speed internet. Music fans were still listening to CD's and using dial-up internet when the 1990's came to a close. Colleges were now offering T1 connections in college dorms, allowing for students to surf the internet and download items at much higher speeds. In 1999, internet developers Shawn Fanning and Sean Parker opened the peer-to-peer file sharing site Napster. Music listeners began sharing their mp3 files, allowing millions of users such as my 18 year old self back in the year 2000 to listen to just about anything available through file sharing.

The record companies hated Napster and the backlash was swift. The record industry was determined to stop Napster and threw all of it's multimillion dollar legal muscle against Napster. But Napster was not without its defenders. Carleton Douglass Ridenhour (Chuck D), the lead vocalist of golden age hip hop group Public Enemy, saw file sharing and live streaming as the opportunity for artists to break free from the tyranny of record companies and their unscrupulous business practices against artists. The record companies prevailed, and A&M Records Vs. Napster ruled that Napster was liable for copyright infringement. The record industry, for many years, resisted the advancement of file streaming technology with all of its might. But resistance proved to be futile.

Paleontologists don't know which exact primate was the ancestor of all of the great apes. But it would have been something like the Proconsul, which appeared during the late Paleogene period. There are many more apes like this in the great ape lineage. They found themselves facing climate induced deforestation. They adapted to the increasing amounts of space in between trees, which contained their sources of food, by learning to walk upright. And over time, their brain sizes of primates would become larger, making them smarter and more able to solve the problems that arise from living in what used to be forest and has now become the savanna.

Early hominids Austalopithicus and Homo habilis learned to make tools. Australopithecus was able to make very simple stone tools, while Homo habilis was able to take this practice much further with the evolution of the opposable thumb. Homo erectus learned to use fire, cook, travel (which required at least very rudimentary navigational skills), and make more sophisticated tools. Traveling in parties required communication, which makes it likely that the evolution of human speech began with Homo erectus.

The Neogene period ended, followed by the Quaternary period (our current period). The Quaternary would see a retreat of the glaciers that formed during the Neogene period, causing the woolly mammoths and saber-tooth cats to die out. Our current species, Homo sapiens, evolved 200,000 years ago during the Quaternary. Homo sapiens began their existence with the hunter-gatherer ways of living of their ancestors. But the major game changer of both our species and the planet was when our species created Agriculture 10,000 years ago.

Agriculture has proven to be the most disruptive advent in natural history since cyanobacteria began pumping the atmosphere with oxygen over 2 billion years ago.

Social Media became a regular influence on people's lives back in the late 2000's. MySpace and Facebook were the titans of the era at the time. MySpace allowed many artists to promote their music. One such artist was Deandre Way (Souljah Boy Tell Em), and he was in the middle of promoting his music. Several internet sensations were going "viral" (a new term for when an internet phenomenon spreaded quickly). Soulja Boy Tell Em would use the internet to gather a huge following, allowing him to get the backing of Collipark Records and Interscope Records to release his hit single "Crank Dat." The song became an instant radio hit and internet sensation.

The song generated mixed reactions from older rappers. Ice-T made comments that went viral, where he criticized the simpleness of the song and said that Soulja Boy Tell Em of "Singlehandedly killed hip hop." He also had his defenders in LL Cool J, as well as people like KRS ONE and Ice Cube that provided more nuanced responses. I personally hated the song and cringed every time it came on. But I appreciated the paradigm shift that was happening. The internet made it impossible for the record industry to fully control who would become a hit and who did not. Since then, there would be several artists that would burst their way into the mainstream using internet promotion.

Music networks such as Spotify began to emerge. Napster returned, and this time legally provided audio streaming music. The record companies had no choice but to embrace this technology. Smartphones began to make the internet even more accessible and eventually would become the primary means of accessing the internet. Older rappers were forced to adapt to the new landscape. Some of their careers were even resurrected by the more egalitarian landscape, where younger audiences were also able to discover hip hop sounds of the past. The record companies still have influence and still find a way to maximize their profits off the successes of artists, such as 360 deals. And the fast moving landscape has made longevity even more difficult for hip hop artists than it already was.

Today, hip hop is a completely unpredictable environment. Mumble rap reigns supreme, but other subgenres are quickly emerging with Snatchat, Tiktok, Facebook, Twitter, and other social media platforms giving artists many different ways to promote content. Montero

Lamar Hill (Lil Nas X) has created a sound that blends the various elements of Atlanta style hip hop with country music, and has taken the country rap subgenre to an unprecedented level. His first singles have already gone diamond since his debut in 2019. Furthermore, his discussing his sexuality and responding to homophobia has pushed the envelope of hip hop social commentary. Fatimah Nyeemah Warner (Noname) from Chicago has a sound that blends Jazz & Neosoul with mellow, highly sophisticated lyrics that discuss a variety of different social and political topics. As of this writing, she is my favorite hip hop artist of this era.

Agriculture is arguably the source of both the best and the worst aspects of Homo sapien existence. It resulted in more time to do things that are not related to survival, which allowed our species to create art, science, music, literature, medicine, and technology. It also created the need for land acquisition, and power, which would lead to class discrimination, sexism, racism, and other forms of human oppression. All of the aforementioned things would become more complicated as human civilization advanced.

It didn't take long for humans to begin slaughtering other species and eliminating them completely from existence. The great auk, the dodo, and Tasmanian tiger are examples of such animals whose extinctions would be accelerated, or completely caused by humans. The issue became more complicated when humans began to develop machines that burned heavy amounts of fossil fuel, releasing enormous amounts of carbon dioxide and other greenhouse gases into the atmosphere. They would also begin destroying forests at an alarming rate, destroying both the habitats of animals and an important means of checking the global carbon cycle.

And today, Homo sapiens are staring at disastrous climate change, an increasing amount of extinctions disrupting the ecosystem, zoonotic viruses infecting humans and causing pandemics such as COVID-19, and other environmental issues that are unintended consequences of human civilization. And despite efforts to curtail environmental destruction, there seems to be no end in sight. Homo sapiens has become a species bent on killing itself on a planet that already has a history of killing 99.9 percent of all living organisms that have ever existed. To make matters more complicated, our enormous brain sizes and sedentary lifestyles have led to an emergence and prevalence of mental illness that has never been seen before in our planet's history.

With the advancement of biotechnology, nanotechnology and AI, I believe that our species will be the first in Earth's history to create its own replacement. We are consciously affecting our own evolution in ways that Homo erectus did not have the capacity to do. We have also become the first species to leave Earth by constructing machines that are powerful enough to escape Earth's gravity. Our species fight for survival may result in leaving the planet, just as our hominid ancestors left the Rift Valley in East Africa.

Hip hop's evolution from rough and unstable beginnings during tough economic times into being a billion dollar industry with an uncertain artistic direction has mirrored our development as a species. We are the descendants of organisms that made it through five major mass extinctions caused by a constantly changing planet. We evolved into organisms whose tool making ability would be both its blessing and its doom.

What happens next for hip hop, humans, life on Earth and the Earth itself? Nothing lasts forever. That much has been certain since the beginning. Earth itself has an expiration date. The sun is bound to swell into a red giant and engulf the planets of the solar system before burning out its remaining fuel and shrinking into a black dwarf. But before that day happens millions of years from now, there will be many more endings and beginnings.

Everything is everything
What is meant to be, will
be
After winter, must come
spring
Change, it comes
eventually
Everything is everything
What is meant to be, will
be
After winter, must come
spring
Change, it comes
eventually

- Lauryn Hill in Everything
is Everything.

We will just have to keep listening to see what's next on the playlist.

Recommended Reading

Brannen, Peter. *The Ends of the World: Volcanic Apocalypses, Lethal Oceans and Our Quest to Understand Earth's Past Mass Extinctions*. Oneworld Publications, 2018.

Brusatte, Stephen. *The Rise and Fall of the Dinosaurs: a New History of Their Lost World*. William Morrow, an Imprint of HarperCollinsPublishers, 2019.

Charnas, Dan. *The Big Payback: the History of the Business of Hip-Hop*. New American Library, 2011.

Clack, Jennifer A. *Gaining Ground: the Origin and Evolution of Tetrapods*. Indiana University Press, 2012.

Dewese, Mohandas "Kool Moe Dee". *There's a God on the Mic: the True 50 Greatest MCs*. Thunder's Mouth Press, 2003.

Edwards, Sue Bradford. *The Evolution of Mammals*. Abdo Publishing, 2019.

Grasse, Tyson Neil de, and Donald Goldsmith. *Origins: Fourteen Billion Years of Cosmic Evolution*. W.W. Norton, 2004.

Griffin, William "Rakim", and Bakari Kitwana. *Sweat the Technique: Revelations on Creativity from the Lyrical Genius*. Amistad, 2020.

Hazen, Robert M. *The Story of Earth: the First 4.5 Billion Years, from Stardust to Living Planet*. Penguin Books, 2013.

Iandoli, Kathy. *God Save the Queens: the Essential History of Women in Hip-Hop*. Dey St., an Imprint of William Morrow, 2020.

Knoll, Andrew H. *A Brief History of Earth: Four Billion Years in Eight Chapters*. Custom House, 2021.

Parker, Derrick, and Matt Diehl. *Notorious C.O.P.: the inside Story of the Tupac, Biggie, and Jam Master Jay Investigations from the NYPD's First "Hip-Hop Cop"*. St. Martin's Griffin, 2007.

Prothero, Donald R. *The Story of the Earth in 25 Rocks: Tales of Important Geological Puzzles and the People Who Solved Them*. Columbia University Press, 2020.

Reese, Eric. *The History of Hip Hop*. CreateSpace Independent Publishing Platform, 2019.

Spitzer, Michael. *The Musical Human: a History of Life on Earth*. Bloomsbury Publishing, 2021.

9 781008 944015

www.ingramcontent.com/pod-product-compliance
Ingram Content Group UK Ltd.
Pitfield, Milton Keynes, MK11 3LW, UK
UKHW041837200726
13854UKWH00003BA/1188